Insulin

Administration

Handbook For

Non-Licensed Personnel

Written By: Robert Palmer, RN, BSN

Copyright © 2013

Course Outline

Overview and Certificate Information

- The course length is 4 hours but may be longer to meet the needs of the class participants or the trainer.
- Class size is limited to 15 students.
- Only a Registered Nurse may teach this course.
- Students must be at least 18 years of age and have a high school diploma or GED.
- Students must attend the entire 4 hours of classroom instruction and participate in discussion and activities.
- The student will need to pass a closed book written exam at the end of the course with a score of at least 90%. If the student fails the exam, he/she will be required to retake the entire 4 hour course.
- The student must complete, to the satisfaction of the instructor, an actual or simulated insulin administration.
- Upon completion of the course, but before actually administering insulin unsupervised, the student must be watched at least twice administering insulin to an individual supported. This does not have to be a licensed nurse, but may be a supervisor or delegated employee. The person watching the insulin administration needs to have at least one year of experience administering insulin and is trusted to follow all of the standards taught within this handbook.
- Once you have completed this course you will obtain a certificate. This certificate is valid for two years. At the end of two years you must retake a two hour refresher course. If you fail to complete a refresher course before your certificate expires, you are not allowed to administer insulin from the date that you are expired until you complete a refresher course.
- If a certificate is expired for more than three months, the student will be required to take the entire four hour class and receive a new certificate.
- If any supervisor within the organization or funding source finds that the person holding an insulin certificate is not safely performing their duties, they shall immediately suspend that person's right to perform such functions. Suspensions may be temporary, but can lead to revocation, depending on the circumstances.
- Revocation of the insulin certificate will occur if personnel do not demonstrate compliance and/or are not performing their duties in a safe manner according to the standards taught in this handbook.
- This course in no way identifies the non-licensed personnel as "Certified". Successful completion of this course allows you to obtain a certificate of completion which must be approved by the organization you are working for along with any State funding source such as the Department of Mental Health.

What is Diabetes?

Diabetes (diabetes mellitus) is when the pancreas cannot secrete enough insulin to meet the demands of the body or the body cannot effectively use the insulin that is produced. The incidence of diabetes in the United States has been increasing rapidly. The earlier diabetes is discovered and treated, the fewer complications the individual will experience.

When an individual eats food, the food is digested and the glucose makes its way into our blood stream. The cells in our body use glucose for energy and growth. However, glucose cannot enter our cells without insulin being present (Insulin makes it possible for our cells to take in the glucose).

Insulin is a hormone and it is produced by the pancreas. Insulin is released as blood sugar levels increase as a response to lower the blood sugar level.

There are three main types of diabetes:

Diabetes Type 1: A person produces very little or no insulin at all.

Diabetes Type 2: A person doesn't produce enough insulin, or the insulin is not working properly.

Gestational Diabetes: A person develops diabetes just during pregnancy.

Many people are surprised when diagnosed with diabetes. Elevated blood sugars occur slowly over a long period of time. Most cases of diabetes are diagnosed when a person goes to a doctor for a routine exam and the doctor notices an elevated blood glucose level on a lab or urine sample. This prompts the physician to order additional testing in order to accurately diagnose the condition.

Symptoms of Diabetes	
* Hunger	* Frequent urination
* Fatigue	* Unexplained weight loss
* Vision changes	* Wounds that don't heal
* Increased thirst	* Frequent infections

Common Causes of Diabetes

➢ Genetic (family history of diabetes)

➢ Gestational (occurs during pregnancy)

➢ Overweight (some people cannot secrete enough insulin to meet the demands of the current body weight)

➢ Autoimmune Disorders (body attacks the pancreas as if it were a foreign to the body)

➢ Medication Side Effects (medication induced diabetes)

➢ Illegal Drug Use (drugs can be hard on the pancreas)

Normal Blood Sugar Ranges

According to the American Diabetes Association a normal blood sugar range is between 70 and 110. Blood sugar testing should occur before a person eats or drinks anything. When a person eats or drinks anything containing calories, the blood sugar level in the blood will raise, generally within three minutes. It is very important that blood sugars are monitored accurately in order for the physician to properly treat the individual.

Hypoglycemia

Hypoglycemia occurs when a person's blood sugar level drops below 70. When this occurs, there is not enough glucose in the blood to provide the energy that the body requires. This can occur for several reasons:

- Taking diabetes medications and did not eat enough food during a meal or snack
- Taking diabetes medications and skipped or delayed a meal
- Took more medications for diabetes than needed for the food eaten
- Exercised or being more active than usual
- Ate or Drank simple sugars without complex carbohydrates or proteins

Signs and symptoms of hypoglycemia:

- Sudden hunger
- Headache
- Shaking
- Fast or pounding heartbeat
- Sweating
- Tired
- Dizziness
- Blurred vision
- Nervousness
- Tired / Fatigued
- Numbness or tingling (mouth and lips)
- Irritability
- Confusion
- Difficulty concentrating
- Slurred or slowed speech
- Unconsciousness
- Seizures

Treatment of Hypoglycemia

If the blood glucose is tested and the reading is below 70 mg/dl and the person is alert and oriented:

1. Administer approximately 15 grams of glucose or equivalent (see chart below).
2. Wait 15 minutes and retest the person's blood glucose level.
3. If the person's blood glucose level is less than 70, administer and additional 15 grams of glucose or equivalent. If the person's blood sugar reading is not above 70 after the second attempt, emergency medical personnel will need to be contacted.
4. If the person's blood sugar is above 70, the person will need to eat a snack containing carbohydrates and protein in order to maintain their elevated blood glucose. Examples include peanut butter and crackers, crackers and cheese, or if it is close to a meal go ahead and let the person eat the meal.

| Hypoglycemic Carbohydrate Examples |
Each item is equal to 15 grams of a fast acting carbohydrate
6 Life Savers chewed and swallowed
8 ounces of milk
6 ounces of juice
Medium sized fruit
1/2 can of non-diet soda
4 glucose tabs
1 tube of glucose gel
1 tablespoon honey, jelly or jam
1 tablespoon pancake syrup
1/2 of a regular candy bar

NOTE: IF A PERSON'S BLOOD SUGAR IS LESS THAN 70 AND THEY ARE UNCONSCIOUS, CONFUSED OR DISORIENTED, YOU CANNOT PUT ANY FOOD OR LIQUID IN THEIR MOUTH. THIS WILL INCREASE THE RISK FOR CHOKING. YOU MUST CALL 9-1-1.

Glucagon Kits:

If a glucagon kit has been prescribed, <u>the glucagon kit may be administered as long as the person administering it has received specialized training</u>. Glucagon should only be administered for a person with a blood sugar less than 70 when the person is either unconscious or it is not safe to put food or liquids in the person's mouth due to the risk of choking. The order for the steps would be (1) Administer the Glucagon (2) Call 911. When glucagon is injected it releases glucose that is stored in the liver in order to raise blood sugar levels quickly.

Administering Glucagon

1. Position the individual on his or her side.
2. Remove the cap from the glass vial.
3. Pull the needle cover off the syringe.
4. Insert the needle into the vial and inject the liquid.
5. Shake to dissolve.
6. Draw the glucagon solution back into the syringe and remove the needle from the vial.
7. Remove air bubbles from the syringe.
8. Insert the needle at a 90 degree angle and inject into a large muscle (upper arm, thigh, buttock).
9. Withdraw the needle and apply slight pressure to the injection site.
10. Keep the individual positioned on their side.
11. Call 911.
12. Remain with the person until Emergency Medical Services assumes control.

Hyperglycemia

Hyperglycemia is defined as a fasting blood glucose level above 110. If a person is diagnosed with diabetes previously, the doctor should have standing orders as to when the person needs to go to the emergency room according to their blood glucose level. As blood sugar levels rise in the blood, the body attempts to get rid of this sugar by making the person thirsty. The goal with the increased thirst is to dilute the sugar in the blood stream. After diluted, the sugar is excreted through the urine.

Signs and Symptoms of Hyperglycemia

- Increased thirst
- Increased urination
- Weakness
- Stomach pains
- Increased respirations
- Loss of appetite
- Nausea and vomiting
- Fatigue
- Confused or disoriented
- Fruity smelling breath

Treatment of Hyperglycemia

If a person has an elevated blood glucose level, they need to be taken to the emergency room if the blood sugar is greater than or equal to the level set by the primary care physician. Typically this level will be around 350 but may be more or less. If the person has oral medications or insulin prescribed, these should be administered as ordered. Additionally, zero calorie drinks may be offered in order to help dilute the sugar in the blood stream and promote urinating out the excess sugar. Exercising, such as walking, should be encouraged to help metabolize as much of the sugar as possible.

Complications of Uncontrolled Diabetes:
• Blindness
• Kidney failure
• Nerve damage
• Amputations
• Wounds that don't heal
• Affects every body organ

Treatment of Diabetes

❖ All types of diabetes are treatable, but there is no known cure. Treatment involves maintaining a normal body weight for the individual, diet and exercise.

❖ Maintaining a normal body weight will decrease the amount of insulin needed compared to being overweight.

❖ Eating a well balanced diet and limiting simple sugars. The doctor should give specific recommendations as to the amount of carbohydrates the person may consume within 24 hours. 180-200 grams of carbohydrates, no more than 25 grams of fat, and 50-60 grams of protein is the generally accepted diabetes recommendations.

❖ Exercise is important. People that are physically fit generally are more efficient in the way their body uses insulin.

❖ Taking medications as prescribed.

❖ Three meals per day plus three snacks per day. In order to maintain blood glucose levels at a steady level without peaks and dips, the person should eat healthy snacks between the major meals each day. The amount of carbohydrates, fat and protein needs to be divided between all foods eaten within the 24 hour period.

Managing Diabetes Through Proper Nutrition

Foods to Avoid

White breads	Cakes / Pies
White rice	Candies
White potatoes	Cookies
White pasta	Donuts
Regular Soda	Juices
Noodles	Chips

Good Food Choices

Whole wheat breads, rice, pastas	Sugar free Jell-O
Low fat popcorn selections	Fruits
Reduced calorie breads	Vegetables
Low fat food selections	Lean Meats
Artificially sweetened desserts	High fiber foods
Diet versions of sodas	Low fat Milk

Good Bedtime Snacks

String Cheese
5 saltine crackers with peanut butter
6 ounce glass of milk
1 oz of nuts (almonds or peanuts)

Daily Requirements

No more than 180-200 carbohydrates per day
Approximately 25 fat grams per day
Approximately 50-60 grams of protein per day
At least 30 grams of fiber per day

There needs to be a good balance between carbohydrates, protein, and fat. Carbohydrates have the greatest effect on a person's blood glucose level, where protein and fat have less of an effect. A person with diabetes should eat three small meals a diet with appropriate snacks between meals. Snacks help to decrease fluctuations in blood sugars by not going too long without food. People with diabetes can eat anything that anyone else can eat, but they need to eat high sugar food in moderation. Always watch portion sizes and read nutrition labels carefully.

Healthy Eating tips

- Do not skip meals
- Eat regularly and space carbohydrates evenly throughout the day
- Meals should include a good source of carbohydrate, lean protein and/or healthy fat
- Include 1 ½ cups vegetables, 2-3 servings of fruit, 1-3 cups of low fat milk or yogurt
- Drink water or zero calorie drinks. Goal is 64 ounces per day (Eight 8 oz glasses)

Foot Care For Diabetics

DO:

- ✓ Wear shoes or slippers at all times
- ✓ Keep skin soft, apply lotion on top and bottom of feet
- ✓ Use only lukewarm water
- ✓ Wear comfortable properly fitting shoes
- ✓ Wear pantyhose or socks with shoes

Don't:

- ✓ Don't go barefoot
- ✓ Don't allow feet to become dry and cracked
- ✓ Don't use hot water
- ✓ Don't wear shoes that don't fit properly
- ✓ Don't wear tight socks or knee highs

Glucometer Training

- A glucometer is a device used to check a person's blood sugar.
- Blood sugar readings should be taken before a person eats or drinks anything.
- Supplies needed:
    - Glucometer Machine
    - Gloves
    - Lancet
    - Sharps Container
    - Alcohol pads
    - Test Strip for the glucometer machine
    - Lancet Pen or other device

Steps:

1. Assemble equipment
2. Identify the individual and explain the procedure
3. Wash hands and apply gloves
4. Place lancet in a lancet pen (if using a device with the lancet)
5. Set up the glucometer
6. Clean individual's finger with an alcohol pad
7. Turn glucometer on, apply lancet to the side of the finger (never the finger pad)
8. Point finger downward and gently squeeze to get an adequate blood sample
9. Place drop on the end of the testing strip and wipe finger with alcohol pad or guaze
10. Read and record the result on the medication administration record
11. Clean equipment and dispose of lancet in sharps container
12. Remove gloves and wash hands

Note:

Some devices allow you to take blood sugar readings from sites other than the fingers. Follow the equipment's recommendations on where blood sugar readings can occur. Always use the fingers unless instructed otherwise.

Rotate fingers to avoid the formation of calluses.

Once a month the machine must be checked using either a "normal" solution or "High and low" solutions. These checks are to verify the machine is working correctly. Follow the recommendations with the solutions and the machine you are using. Always document the "normal" reading or "highs and lows" on the MAR. Do not use the machine if the controls are not within their correct ranges.

Some machines require you to calibrate the machine every time a new box or bottle of test strips is opened. Always read the information that comes with your machine and follow all directions in order for the machine to give accurate results.

Oral Diabetic Medications

Trade Name	Generic Name	How it Works	Positive Effects	Negative Effects
Glycoset Precose		Blocks the absorption of carbohydrates in the intestines	Lowers blood glucose after meals. Does not cause hypoglycemia	May cause bloating, gas, and diarrhea
Glucophage	Metformin	Helps the body use its own insulin better	Does not cause hypoglycemia. May help with weight loss and improve cholesterol levels	May cause diarrhea. Avoid alcohol.
Prandin		Lowers blood glucose by stimulating the pancreas to release more insulin	Absorbed quickly. Used alone or with Metformin.	Risk of hypoglycemia if receiving too much medication compared to food eaten
Actos Avandia	Pioglitazone Rosiglitazone	Improves the muscles ability to use insulin. Does not make more insulin.	Won't cause hypoglycemia if taken by itself	May cause liver problems. May interfere with birth control pills effectiveness
Diabeta Glucotrol XL Amaryl Micronase	Glyburide Glipizide Glimepiride Glynase	Stimulates the pancreas to produce more insulin	Fewer side effects compared to other medications	Risk of hypoglycemia if receiving too much medication compared to food eaten

Types of Insulin

Insulin Type	Brand Name	Onset	Peak	Duration	Appearance	Can be Mixed with
Insulin aspart	NovoLog	5-10 min	1-3 hrs	3-5 hrs	clear	NPH but only with doctor's approval. Mix and inject immediately
Insulin glulisine	Apidra	15 min	30 min - 1 hr	3-5 hrs	clear	NPH
Insulin lispro	Humalog	15 min	30 min - 1 hr	4-5 hrs	clear	NPH, Ultralente with doctor's approval. Lispro drawn up 1st, mix and inject immed.
Regular	Humulin R Novolin R	30 min - 1 hr	2-4 hrs	5-7 hrs	clear	NPH, Lente, Ultralente
Lente		1-3 hrs	6-14 hrs	24+ hrs	cloudy	Regular
NPH	Humulin N Novolin N	1-2 hrs	6-14 hrs	24+ hrs	cloudy	Regular
Insulin detemir	Levemir	1 hr	"peakless"	up to 24 hrs	clear	Do not mix
Insulin glargine	Lantus	1 hr	"peakless"	24 hrs	clear	Do not mix
Ultralente		6 hrs	18-24 hrs	36+ hrs	cloudy	Regular
Insulin aspart protamine suspension/aspart	Novolog Mix 70/30	10-20 min	1-3 hrs	up to 24 hrs	cloudy	Do not mix
Insulin lispro protamine suspension/lispro	Humalog Mix 75/25	15-30 min	30 min - 2 hrs	14-24 hrs	cloudy	Do not mix
NPH suspension/ Regular	Humulin 50/50	30 min - 1 hr	2-5 hrs	14-24 hrs	cloudy	Do not mix
NPH suspension/ Regular	Humulin 70/30 Novolin 70/30	30-60 min	2-6 hrs	14-24 hrs	cloudy	Do not mix

All insulin listed above are stable at room temperature for 28 days in the original vial. For insulin pens and similar delivery devices, see package inserts for the amount of time the pen or device may be stored at room temperature. All of the action times shown are estimates. The onset, peak, and duration of action may vary due to individual variations.

Opening and Labeling Insulin Vials

- ✓ Once the cap is removed from an insulin vial, you must initial and date the vial using permanent ink.
- ✓ The vial of insulin must be discarded after being open 28 days.

Insulin Storage

1. Never freeze (frozen insulin needs to be discarded).
2. Refrigerated Insulin should be kept between 36-46 degrees F.
3. Never use insulin past the expiration date stamped on the vial, pen, or cartridge supplied from the drug manufacturer.
4. After an insulin bottle's cap has been removed the insulin may only be used for up to 28 days before it must be discarded. Always follow the manufacturer's instructions.
5. Never expose the insulin to direct heat or sunlight.
6. Inspect insulin prior to each use. Any insulin that has clumps or solid white particles should not be used. Insulin that is supposed to be clear should not have any cloudy appearance.
7. Check storage guidelines specific to the insulin being used.
8. Opened insulin currently being used can generally be left at room temperature 59- 86 degrees F.
9. When storing pre-filled insulin syringes, store them with the needle pointing up. Most insulin can be stored in a syringe (Single formulation) for up to 30 days refrigerated, but always refer to the package insert for specific storage information. Certain insulin types such as Lantus and Novolog cannot be pre-drawn and stored.

Injection sites:

Insulin is injected into the fatty area (subcutaneous) under the skin and above the muscles. After you inject insulin, it is absorbed into the blood where your body can distribute it. The safest locations to inject insulin include:

- ✓ Abdomen (at least one inch away from the belly button)
- ✓ Back of the upper arms
- ✓ Hip area
- ✓ Top part of the thighs

Angle of injection:

Insulin is generally injected at a 90 degree angle. If a person is thin, the injection may be given at an angle between 45 and 90 degrees.

Rotating Injection Sites:

- Inject your insulin in the same general area for 1-2 weeks. Each time you inject in that area, put the needle in a different spot. At the end of 1-2 weeks, move to another area of the body.
- If you are very thin or very muscular, you may need to avoid using your arms or legs.
- It is not a good idea to use only your arms or legs. There is not enough room to move around in these areas.
- If using only the abdomen, inject in a new spot each time.

Proper Disposal of Needles, Syringes, and Lancets

- o Place syringes, needles, and lancets in an approved sharps container. Never throw loose syringes or lancets in a trash can.
- o Keep sharps container in an area that is safe, out of the reach of children and individuals supported who may place their hands inside of the container.
- o When the container has been filled to the appropriate level, the container needs to be secured and disposed of properly. Containers generally have maximum fill lines marked on the containers. Containers should only be filled up to ¾ full. Sharps container filled past the ¾ mark increase the risk for injury from a needle stick.
- o Filled containers cannot be placed in recycling bins; you cannot recycle syringes or lancets.
- o Check with your town's trash or removal company for safe disposal of used syringes and lancets. The local health department is also a good resource for sharp container disposal information.

Drawing Up and Injecting Insulin

1. **Gather Supplies:** Insulin syringe, alcohol pad, gloves, insulin, and sharps container

2. Wash hands and put on disposable gloves

3. Roll bottle between palms of your hands (never shake insulin)

4. Wipe the top of the insulin bottle with an alcohol pad.

5. Inject ___ units of air into the insulin bottle, do not inject the air into the insulin itself (bottle should be upright)

6. Turn the bottle upside down with the needle still in the bottle and remove ___ units of insulin

7. Make sure there are no air bubbles in the syringe

8. Pick injection site and wipe the area with an alcohol pad. Let the area air dry.

9. Pinch up the skin, insert the needle and push the plunger in, then remove the needle

10. Place the needle and syringe in an approved sharps container

* Always verify you have the correct person, insulin, dose, and time
* Always verify that the insulin is not expired and has been stored correctly
* Insulin is ordered in units only. You can only use an insulin syringe calibrated in units
* If insulin is ordered before a meal, the insulin should be administered anywhere from 15-30 minutes before a person eats. Always read the product information as follow their guidelines.

Important information Regarding Insulin Administration

➢ The only time you can recap a needle is when you prepare the insulin in advance and are not ready to administer the injection immediately. In order to recap a needle you place the cap on a hard, clean surface and gently guide the needle into the cap. Once the cap is over the needle you can use your fingers to secure the cap in place. Never hold the cap in your hand and attempt to recap the needle.

➢ **Never recap a needle after you have injected insulin into a person.**

➢ If at any point you are stuck by a needle after it was used to inject insulin into a person, you must immediately notify your supervisor.

➢ Make sure all air is out of the syringe before administering insulin. If air is taking up space in the syringe then you will not be administering the correct dose of insulin. It is not safe to administer air via injection into a person.

➢ When injecting air into an insulin bottle, make sure the bottle is standing up correctly and the air is administered into the top portion of the bottle and not directly into the insulin. If injected into the insulin, this will create air bubbles.

➢ When administering the actual injection, hold the syringe like a dart and inject at a fairly fast rate at the appropriate angle. Slower injections are uncomfortable.

➢ Do not rub the area after the injection. This will speed up the rate of absorption and may cause the area to bruise or discolor. You may gently wipe the area with the alcohol pad. If the area starts to bleed it is best to apply gentle pressure with a piece of 2x2 gauze without rubbing.

➢ Avoid injecting insulin into the skin where a person has a bruise, or any other places where the skin is not intact and healthy in appearance.

➢ Always use an approved insulin syringe. Insulin is measured in units and not cubic centimeters (cc) or milliliters (ml). 100 units of insulin is actually equivalent to 1 cc.

➢ Never reuse needles in an ISL or Group Home setting. This increases the risk of injury from a needle stick. Needles become dull after just one use. Repeated use will also cause more discomfort to the individual receiving the injection.

➢ Use the appropriate size insulin needle according to the amount of insulin ordered.
   o If injecting less than 30 units use a 3/10 ml/cc insulin syringe.
   o If injecting 30 units to 50 units use a ½ ml/cc insulin syringe.
   o If injecting 51 units to 100 units use a 1ml/cc insulin syringe.

Mixing Insulins

1. Roll the bottle of cloudy insulin between your palms

2. Wipe the top of <u>BOTH</u> insulin bottles with an alcohol pad

3. Insert _____ units of air into the cloudy insulin bottle, then remove the needle

4. Insert _____ units of air into the clear insulin bottle

5. Leave the needle in the clear insulin bottle and remove, turn the bottle upside down and remove ____ units

6. Carefully insert the needle into the cloudy insulin bottle and withdraw ____ units of insulin.

* Make sure all air is removed from the syringe after withdrawing the cloudy insulin
* Make sure that the cloudy insulin does not go into the clear insulin vial
* You can only draw out the number of units ordered for the clear insulin, if you draw out too much, you have to start over with a new syringe.

How to Use an Insulin Pen

An insulin pen is comprised of a vial of insulin and a disposable needle. The pen looks like a large marker. You should always review the proper way to use the prescribed pen with a doctor. The information included here will give you a general idea of how to use a prefilled insulin pen.

Supplies Needed:

- 2 alcohol pads
- Prefilled insulin pen
- New needle for the pen
- Sharps container

Step 1: Select an injection site

The most common sites are the back of the upper arms, top of the thighs, hip region, or abdomen.

Step 2: Protect yourself

Wash your hands, wear disposable gloves and then clean the area selected for the injection with an alcohol pad and allow to air dry.

Step 3: Check the insulin

Carefully check the insulin to make sure it is the correct consistency, color, and not expired. If the insulin is a cloudy insulin, gently roll the pen between your palms to mix it (never shake it).

Step 4: Attach the needle

Always use a new needle. Remove the protective paper tab from the outer needle shield and attach the shielded needle straight onto the pen. Save the shields for when you detach the needle.

Step 5: Prime the pen

Turn the dose knob to the arrow. Pull the knob out until a "0" appears in the dose window. Turn the dose knob clockwise until a "2" appears in the window. With the needle pointed up, tap the cartridge holder gently. Push the injection button firmly. You should see several drops of insulin coming from the needle tip.

Step 6: Give a dose of insulin

Return the dose knob to zero, then turn the dose knob to the number of units you need to administer. Pinch the skin and fat to make sure you are administering the insulin into fatty tissue. Using a 90 degree angle insert the needle, release the tension on the skin and depress the plunger. Count slowly to five and remove the needle.

Step 7: Clean up

Gently wipe the insertion site with an alcohol pad. Attach the outer needle shield to the pen then unscrew the needle. Dispose of the needle in an approved sharps container.

Insulin Pumps

- Insulin pumps are primarily used for people with type 1 diabetes.

- Insulin pumps deliver short acting insulin all day and night.

- The pump looks like a small pager and delivers insulin through a catheter placed under the skin.

- Basal insulin is the normal level of blood insulin when you have not eaten or when you are asleep. Basal insulin is delivered constantly throughout the day and night.

- Bolus insulin (extra insulin). When you eat your blood will need more insulin. You press buttons on the pump which will give additional insulin. If your blood glucose is too high you can administer a bolus dose to bring it back down.

- Most people attach the pump to a waistband using a clip or case similar to a pager. The pump may also be placed inside of a person's pocket.

- Actual pump instructions vary according to the device.

- Advantages include: no more injections, the pump is accurate, and better overall control of blood glucose levels.

Sample Documentation on an MAR

Insulin Regular 10 units subcutaneous BID	8:00 AM	rl	rl	ll	ll	rl	lj	jl	jl	pl									
	5:00 PM	MP	BL	MP	BL	MP	BL	MP	BL	LP									
Check Blood Sugar 8am, 5pm before meals	8:00 AM	136	142	128	99	106	160	180	110	106									
		MP	BL	MP	BL	MP	BL	MP	BL	LP									
	5:00 PM	142	150	150	152	148	160	148	140	142									
		MP	BL	MP	BL	MP	BL	MP	BL	LP									
Monthly High and Low Check	High				223														
					RP														
					RP														
	Low				67														
Blood Sugar Check Q HS																			
		258	300	322	328	198	226	148	140	348									
	8:00 PM	MP	BL	MP	BL	MP	BL	MP	BL	LP									
Regular Insulin subcutaneous Q HS IS BS 200-250 give 2 units, 251-300 give 3 units, 301-350 4 units, >351 call doctor																			
		3 u	3u	4u	4u	0u	2u	0 u	0u	4u									
	8:00 PM	RP	RP	LB	LB	KP	RP	LB	KP	RP									
Blood Sugar check PRN if signs of Symptoms of Hyperglycemia or Hypoglycemia	PRN																		

45

Student Name _____

Date _____

Skills Check Sheet

	Yes	No	Comments
Student washed their hands			
Used Disposable Gloves			
Gathered all necessary supplies			
Verified correct person, insulin, dose, and time			
Verified the Insulin was not expired or opened past 28 days			
Cleaned the top of the insulin bottle with an alcohol pad			
Injected the correct number of units of air into the insulin bottle and not into the insulin itself			
Withdrew the correct number of units of insulin with no air bubbles in the syringe			
Administers the insulin in an approved location			
Insulin administered using a 90 degree angle or as recommended by the instructor			
Needle and syringe was discarded in an approved sharps container			
Needle was **not** recapped after the injection			

Pass _____   Fail _____

Instructor Signature _____ Date _____

Name _____   Date of Training _____

Date of Birth _____   Social Security Number _____ - ____ - _____

Agency Working For _____

Training included each of the following topics:

- Overview of the course and certification
- Definition of Diabetes
- What causes diabetes (genetics, viruses, drugs, Immune response, Pregnancy)
- Normal Blood Sugar Ranges
- Hyperglycemia (Overview)
- Hypoglycemia (Overview)
- Diabetes Mellitus I and II
- How to treat Diabetes Mellitus I and II
- Education, Diet, Exercise for people with Diabetes
- Types of Oral Medications for Diabetes
- Types of Insulin for Diabetes
- Glucometer training (blood sugar testing)
- Calibration of Glucometer machines
- Monthly testing of highs and lows (glucometer)
- Types of devises for administering Insulin
- Storage of Insulin
- Opening bottles of insulin and labeling them
- Documentation of Insulin Injections
- Where you can inject insulin
- Drawing up insulin into a syringe
- Importance of rotating injection sites
- Sliding scale Insulin orders
- Complications of untreated diabetes or noncompliance
- Glucagon Injections (if ordered)
- Practice session (drawing up insulin and injecting into a injection pad)
- Completing paperwork for certificates

Student Signature _____

Diabetes / Insulin Administration

Name _____

Date _____

(1) What is Diabetes?

_____

_____

_____

(2) Outside of Medication how is Diabetes Managed?

_____

_____

(3) In regards to Diet, what foods should be encouraged and what foods should be used sparingly?

_____

_____

_____

(4) What are some examples of healthy snacks a diabetic can eat?

_____

_____

(5) If a person's blood sugar is low (less than 70) and they have tremors and obvious signs of Hypoglycemia, but they are alert and oriented, how would you help them get their blood sugar level up?

_____

_____

_____

_____

(6) Name two signs and symptoms of hypoglycemia (Fasting blood sugar below 70).

_____

_____

(7) Name two Signs and symptoms of Hyperglycemia (Fasting blood sugar above 110).

_____

_____

(8) How long is Insulin good for once the cap is removed and bottle is initialed and dated?

_____

(9) Name two sites where insulin can be administered (be specific)?

_____

_____

(10) Where do you dispose of your needle once the insulin has been administered?

_____

_____

(11) True or False (Circle one) Always inject the number of units of air into the insulin bottle before trying to draw out the units of insulin required?

(12) What would you do for a person that had a blood sugar reading of 30 and you cannot wake him up?

_____

_____

_____

Works Cited

American Diabetes Association, Diabetes Pro, Professional Resources On-line.
http://professional.diabetes.org

A Practical Guide to Clinical Medicine. University of California, San Diego. 2011.

Nursing 2011 Drug Handbook. Lippincott, Williams & Wilkins. 31st Edition.

Flex Pen Insulin Delivery System. August 2011 Novo Nordisk.
http://myflexpen.com

"How Insulin Pumping Works." Animas Corporation 2007 – 2010. Updated November 22, 2011. http://animas.com